ROURKE
SCIENCE
PROJECTS

SCIENCE IN ART

Authors: George and Shirley Coulter

Rourke Publications, Inc.
Vero Beach, Florida 32964

A book by Market Square Communications Incorporated
Pamela J.P. Schroeder, Editor

LIBRARY OF CONGRESS CATALOGING-IN-PUBLICATION DATA

Coulter, George, 1934-
    Science in art / by George and Shirley Coulter.
        p.   cm. — (Rourke science projects)
    Includes index.
    ISBN 0-86625-519-2
    1. Science—Experiments—Juvenile literature. 2. Science projects—Juvenile literature.
3. Art and science—Juvenile literature. [1. Science projects. 2. Science—Experiments.
3. Experiments. 4. Art and science.] I. Coulter, Shirley, 1936-   . II. Title. III. Series.
Q164.C677   1995
507.8—dc20   94-49347
                                                                              CIP
                                                                               AC

Printed in the USA

TABLE OF CONTENTS

PICTURE YOURSELF AS A SCIENTIST

Do you like to ask questions? Then you already have the makings of a real scientist!

Scientists ask questions about why things are the way they are, and then they search and test for the answers. Inside this book, you'll find questions about SCIENCE IN ART. Choose one—or more—that you want to investigate.

After scientists choose a question, they sometimes try to guess the answer, based on their experience. That guess is called a **hypothesis** (hii POTH uh siss). Then they experiment to find out if their hypothesis is right.

Once you choose your question, you'll start to experiment, using the steps written out for you. You'll be acting like a professional scientist, making

Scientists and artists alike are careful observers. They use what they see, hear, taste, smell and feel to understand the world around them.

careful **observations** (ahb zer VAY shunz), and writing down all your results in a **science log** (SII ens LAWG). Your notes are very important. You'll need to use them to make a display to share what you've learned with other people.

Please be careful while you're experimenting. Professional scientists are always aware of safety.

At the end of your experiment, you'll find—answers! Other people will believe your answers because you have scientific proof. However, you don't have to stop there. Your answer might lead to another question. Or you might want to find out about something else. Don't wait. Just picture yourself as a scientist—then do it!

SCIENCE IN ART

Science is all around us. It's not just a separate subject you learn about in school. It's everywhere—in the air you breathe, the food you eat and the pictures you paint.

Just think of anything you might need to draw a picture, or paint a painting. Where does color come from? What makes paint, or chalk or a pencil draw? Why is light so important to artists? These are all questions that have to do with science.

When mobiles move, they follow **physical laws** (FIZ i kul LAWZ). **Dyes** (DIIZ) and paints use the science of **chemistry** (KEM iss tree) to spread their colors. Light—**refracted** (ree FRAKT ed) and **reflected** (ree FLEKT ed)—allow us to see shapes and colors.

Scientists and artists work hand in hand. In their laboratories, scientists create new materials and technologies that artists use in their work. Artists create computer programs that help scientists to see their ideas better. They also make the drawings that appear in your science textbooks.

So the next time you take out your paints, make a collage or create a sculpture, remember—science is a part of you and all you do.

WHAT COLOR IS BLACK INK?

Everyone has a favorite color. People often like to buy and wear clothes that are dyed their favorite color.

However, did you know that some colors hide other colors inside? Understanding the science of color—chromatics—means artists know what colors are really made of.

What You Need

✓ large, round coffee filters
✓ pressing iron
✓ scissors
✓ straight pin
✓ set of water-soluble felt-tip pens (make sure to have one black pen)
✓ one or more bowls
✓ tap water
✓ white paper toweling

What To Do

Step 1 Ask an adult to help you use a pressing iron to flatten at least 10 coffee filters. Be careful with the hot iron.

Use your black pen to make a solid dot in the center of a filter—no more than 1/4 inch (5 mm) in **diameter** (dii A mi tuhr). Make a hole in the center of the dot with the straight pin—about 1/8 inch (3 mm) in diameter.

Step 2 You're going to make a wick to put in the hole you just made. Cut a strip from another filter about 3/4 inch (2 cm) wide and 2 inches (5 cm) long. Twist the strip until it looks like a giant candle wick.

Gently push the wick through the hole in the black dot until it just comes through the other side. Almost all of the wick—2 inches (5 cm)—should hang from one side.

Step 3 Fill a bowl until the tap water is 1 inch (2.5 cm) from its top. (Your bowl should be smaller in diameter than your filters.) Make sure to keep the rim dry. Center the filter, wick down, over the bowl. Lower the filter onto the rim of the bowl—only the wick should touch the water—and wait. Write down your **observations** in your **science log.**

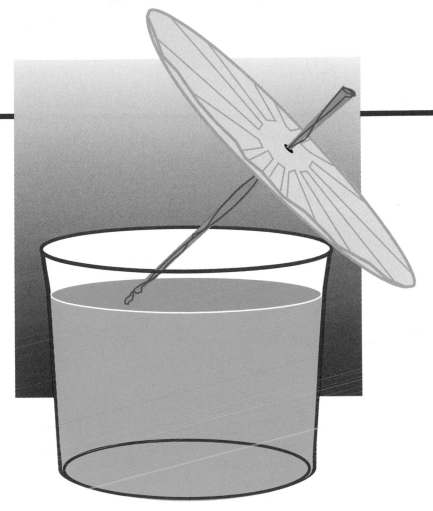

Place your filter, wick down, into the bowl of water. Make sure that only the wick touches the water.

Step 4 After 15 to 20 minutes remove the filter paper from the bowl and set it on white paper toweling to dry. Save your filter paper for your final display.

Step 5 Repeat Step 1 through Step 4 *except* instead of a black dot, make a green dot, a red dot, an orange dot. Use any colors, or combination of colors you have. Remember to record your observations for each color. You may want to label your filters with the color of dot you made at the center.

Step 6 Repeat Step 1 through Step 4 *except* instead of making one black dot in the center of your filter, make four black dots equally spaced—about 3/4 inch (2 cm)—from the center. Make a central hole for the wick, just as in Step 2. Does anything different happen? Make sure to write it down.

Try using more than four dots, or using different colors for each dot.

Is This What Happened?

Step 3: As the water soaked up the wick and through the filter, different colored rings should have appeared, coming from the center outward. The black dot should have created yellow, red and blue circles. Only red should have spread out from the red dot—and the same for blue or yellow. However, a green dot should make yellow and blue rings, orange should make yellow and red, and purple should make—you probably guessed it—red and blue.

Step 6: With four black dots you should have seen four bands of color (each including yellow, red and blue) going out from the center of the filter like rays. More dots would create more bands of color.

Why?

Inks are made from colored substances called **dyes.** Although you can't see them, red, yellow and blue dyes are all needed to make black ink. Red, yellow and blue are primary colors. By mixing these primary colors, you can make all the other colors. What dyes do you think are used to make green ink?

The process of **chromatography** (krohm uh TAH gruh fee) separates all the dyes used to make one color of ink. Some dyes are more **soluble** (SOL yoo buhl) than others. That means some dissolve more easily into water. As the water moves up the wick and through the ink, the dyes that are more soluble move more easily with the water—and end up farthest away! Those that are less soluble move more slowly and show up closer to the center. Which is more soluble—yellow or blue dye?

Artists mix science and art together when they mix paints on their palettes. The science of colors and how they work together blends with the artist's imagination to create a fantastic picture.

Your ink dots may have separated on your filters into colors like these. Understanding where colors come from and how they work together helps artists to create beautiful works of art.

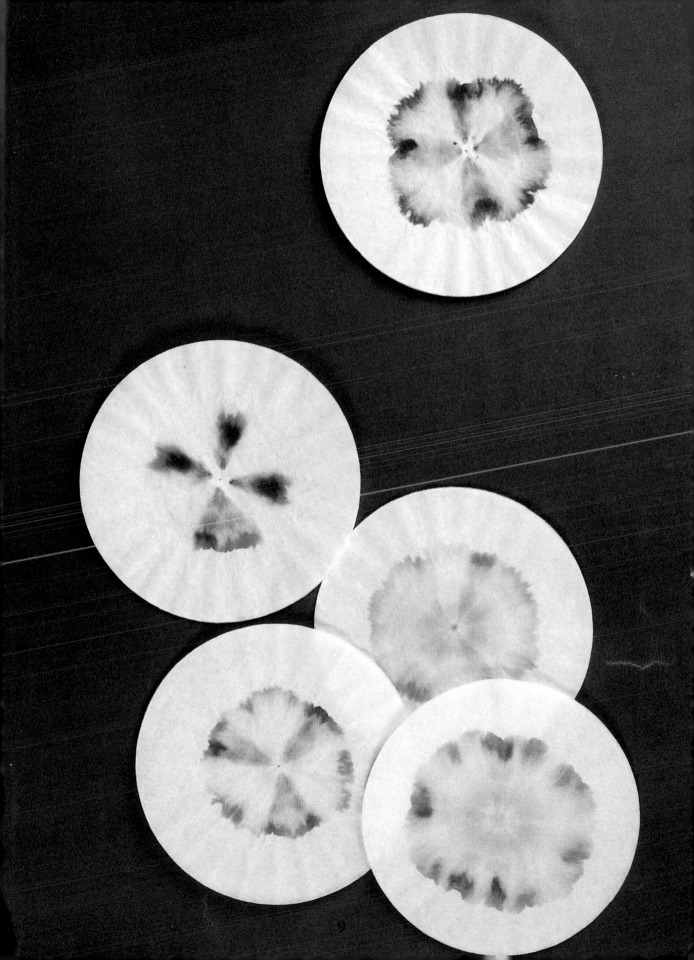

HOW DO ARTISTS MAKE MOBILES BALANCE?

Mobiles—the first one you saw may have hung over your crib when you were a baby. They seem simple, don't they?

Hidden within the beauty and grace of a whirling, twirling mobile lies the **principle** (PRIN sih puhl) of **levers** (LEH vuhrz).

What To Do

Step 1 Have an adult help you use a sharp knife to cut the wood rod, or **dowel** (DOW uhl), into two pieces—one 16 inches (40.6 cm) long, and one 8 inches (20.3 cm) long.

Step 2 Use your compass to draw three circles on posterboard—a 9-inch (23-cm) circle, and two 3-inch (7.6-cm) circles. (You could use different colors of posterboard to make your mobile look more interesting.)

With the sharp end of your compass, make a hole 1/2 inch (1.3 cm) from the rim of each circle—just wide enough to let a piece of thread through.

What You Need

✓ posterboard or other cardboard
✓ a drawing compass with pencil
✓ scissors
✓ 1/8 inch (3 mm) by 3 feet (91 cm) wood rod, or dowel (available at hardware stores)
✓ sharp knife
✓ thread
✓ white glue

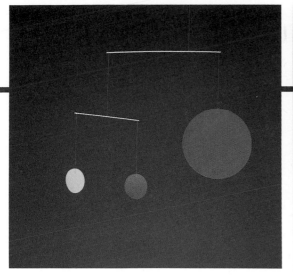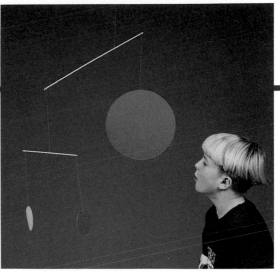

Left alone, mobiles hang in balance. Touching or blowing on a mobile can move it out of balance.

Step 3 Cut a piece of thread about 10 inches (25 cm) long. Tie one end to the *large* circle through the hole you made. Tie the other end around the *long* wood rod, 1/8 inch (3 mm) from the end. Then glue the thread into place.

Cut two more 10-inch (25-cm) pieces of thread. Tie one thread to each of the *small* circles. Then, tie one small circle about 1/8 inch (3 mm) from the end of the *short* wood rod. Tie the other small circle to the opposite end of the short rod, about 1/8 inch (3 mm) from the end. Glue the threads into place.

Step 4 Cut another 10-inch (25-cm) piece of thread. Tie one end in the middle of the *small* rod. Lift the rod by holding only the loose end of the thread. Does it balance? Move the center thread until both small circles hang at the same level. Then glue the center thread into place.

Tie the free end of the thread about 1/8 inch (3 mm) from the end of the *long* rod—opposite the large circle, and glue into place.

Step 5 Cut a piece of thread 12 inches (30 cm) long. Measure 4 inches from the end of the long rod where the large circle is attached, and tie the thread at that spot. Now, test the balance by lifting the free end of the thread until your mobile is hanging free. If one end is lower than the other, move the thread bit-by-bit toward the end that dips. When everything is level, glue the thread into place.

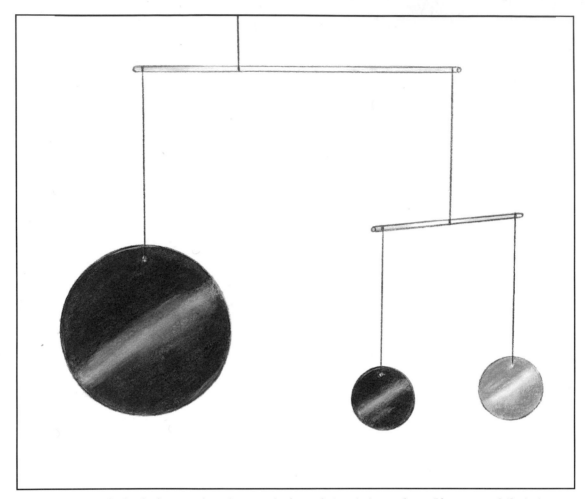

Be sure to check the balance of each piece before gluing it into place. If your mobile is in balance, it should look something like this one when it is finished.

Step 6 Find a place to hang your mobile. Gently try to push the mobile out of balance. Push and pull the circles. Blow at it. Push the ends of the rods up and down. You can even try to add pieces to your mobile. What happens? Write down your **observations** in your **science log.** Take photographs to use in your final display.

Is This What Happened?

Step 6: Each time you disturbed the mobile, it should have come back into balance. When you blew on it, it should have moved in the direction you blew. When you pushed on it, it should have moved in the direction you pushed. When you pushed up or down on the end of a rod, it should have bounced until it regained its balance.

Why?

Mobiles actually have a lot in common with seesaws! Have you ever tried to balance two people on the ends of a seesaw? It only works if both people weigh the same—or if the heavier person sits closer to the middle. Why is that?

Seesaws—and mobiles—are *first class levers.* A first class lever has a **fulcrum** (FUL kruhm) between its two ends. The fulcrum is the balancing point of the lever, and the lever turns on the fulcrum. The fulcrum on a seesaw is the bar the seesaw rests on. How many fulcrums does your mobile have?

As first class levers, seesaws and mobiles must follow the *law of the lever*—an object will balance if its weight times its distance from the fulcrum equals the weight times the distance from the fulcrum of the object on the opposite side. Heavier objects need to be closer to the fulcrum to balance with a lighter object. Making mobiles is a delicate balance of art and science.

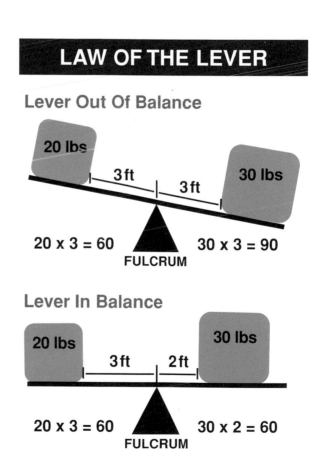

LAW OF THE LEVER

Lever Out Of Balance

20 lbs

3 ft

3 ft

30 lbs

20 x 3 = 60   30 x 3 = 90

FULCRUM

Lever In Balance

20 lbs

3 ft   2 ft

30 lbs

20 x 3 = 60   30 x 2 = 60

FULCRUM

An object will balance if its weight times its distance from the fulcrum equals the weight times the distance from the fulcrum of the object on the opposite side.

HOW CAN YOU USE LIGHT TO CREATE KALEIDOSCOPE PATTERNS?

Have you ever looked through a kaleidoscope? Or watched fantastic patterns revolve and change as you turned the tube?

Kaleidoscopes use light to make the designs you see. Without the science of light and reflection, kaleidoscopes would be awfully dim and dull.

What To Do

Step 1 Use your scissors to cut the curved rim off the plate. You only want to use the flat bottom part. Use a felt-tip pen and ruler to draw three rectangles on the plate—6 inches (18 cm) long by 1 inch (2.5 cm) wide. Cut out your rectangles and you've just created three small mirrors!

Step 2 Place your rectangles, shiny side down, with the long edges next to each other. Tape the long edges together on the less shiny side. Then bring the last two long edges together to form a triangular tube. (Make sure the shiny surface is inside the tube.) Tape these edges together.

You may have to trim the top and bottom of your tube so that all the edges are even. Finally, cover the outside of your tube with dark, colored paper, or a layer of tape. (This will keep light from leaking in the sides of your tube.)

What You Need

- ✓ 12-inch silver, plastic party plate you can see your face in (available at stores that sell party favors)
- ✓ scissors
- ✓ ruler
- ✓ black felt-tip pen
- ✓ masking tape
- ✓ clear plastic wrap
- ✓ colored plastic wrap
- ✓ construction paper or tagboard
- ✓ tracing paper
- ✓ white glue

The science of light is very important to artists. The way light reflects—as in this kaleidoscope—helps us identify colors and shapes, both in life and art.

Step 3 Cut a triangular piece of clear plastic wrap about 1/2 inch (1.3 cm) larger than the end of your tube. Cover one end with your plastic triangle and tape it down on the sides of the tube. This will be your window.

Step 4 Cut a strip of tagboard 1 inch (2.5 cm) wide by 4 inches (10 cm) long. Bend the long edge of the strip around the outside of your window to form a triangle. This triangle "cap" should fit over your window, but not so tightly that you can't take it off.

Take the cap off the window and glue the short edges of your tagboard strip together to keep the triangle shape. Glue a piece of tracing paper over one end of the triangle cap, and set it aside to dry.

Once you've put all the pieces of your kaleidoscope together, fill the cap with small pieces of material and push the tube's window into the cap.

Step 5 Cut up small pieces of colored plastic wrap or anything else that is **transparent** (tranz PAYR ent)—that you can see through. Once the glue on your cap is dry, place these pieces in the triangle cap.

Step 6 Hold the cap so the small pieces will not fall out. Then, gently push the window of your tube into the cap, but not all the way. Leave enough room so that the pieces can move under the window.

And—you've made a kaleidoscope! Hold the cap/window end toward the light and look through. What do you see? Shake or turn the tube. Write your **observations** down in your **science log.** Try to sketch some of the patterns you see.

Step 7 Remove the transparent pieces. Repeat Step 5 and Step 6 *except* use small pieces of cardboard, paper or anything that is **opaque** (oh PAYK)—that you can't see through. What's different? Record your observations.

Is This What Happened?

Step 6: When you looked through your kaleidoscope, you should have seen a six-sided shape with repeating patterns. You should have seen more than one set of patterns, and all parts of the patterns should have been colored. When you shook or turned your kaleidoscope, the patterns should have changed.

Step 7: You should see the same things as in Step 6, *except* the patterns should have some parts without color.

Why?

When you held the kaleidoscope up to the light, the light passed through the tracing paper and was scattered. That's why the background looks white. The colored transparent pieces blocked some of the light, but not all. Red transparent pieces let only red light through, blue let only blue light through, and so on. The opaque pieces completely blocked the light, so they looked black.

Why do the patterns repeat? They repeat because of the mirrors inside your tube. They **reflected** the image that was formed in your cap. You actually saw one real image and several reflections of that image.

The number and kind of designs a kaleidoscope makes is limited only by your imagination and the materials you use. Kaleidoscope patterns can truly change as fast as the speed of light!

The light passes through your kaleidoscope's cap and reflects off the mirrors to create a pattern. Only one of the images you see is the real thing. All the others are reflections.

HOW CAN YOU CREATE COLORS WITH A GLASS OF WATER?

Sometimes after a rainstorm, you can run outside to watch a rainbow appear. Artists are interested in rainbows not only because they are beautiful, but because of the colors they produce. Where do those colors come from?

Have you ever heard the story of how a rainbow leads to a pot of gold? There might be a way to test this idea right at home!

What You Need

- ✓ clear drinking glass
- ✓ tap water
- ✓ 6-volt, large flashlight
  (must produce a broad, bright white beam)
- ✓ prism
  (check with your teacher or school, or available at toy and hobby stores)

What To Do

Step 1 Fill a glass about 3/4 full of water. Place the glass on the edge of a table in a dimly lit room. Wait for the water to be still in the glass.

Step 2 Turn on your flashlight and hold it about 1 foot (30.5 cm) away from the glass. The beam of light should be just about level with the surface of the water. Then, tilt the flashlight so that the light beam hits the water slightly from above.

Step 3 Watch where the light hits the floor in front of the glass for your results. If you place a white piece of paper where the light hits the floor, you may be able to see better. (Don't worry if nothing happens the first time. You may have to try this several times to get your flashlight angle just right, before any results appear.) Record your **observations** in your **science log.** You may want to make a drawing of your setup and results to include in your final display.

Step 4 Set a **prism** (PRIZM) on its base—the long way—on the edge of the table. Hold the flashlight against the opposite edge of the table, so only the top half of the light beam reaches the prism. Look on the floor in front of the prism. What do you see? Be sure to write down your observations.

Step 5 Try to use sunlight as a substitute for the flashlight. You will have to move the glass of water and prism around in front of a window to get any results. Do you see any differences? Keep a record in your science log.

The angle of the light has to be just right to make the spectrum appear. Look on the floor next to the table for any results.

Is This What Happened?

Step 3: You should have seen a curved **spectrum** (SPEK trum)—bands of colored light—very much like a mini-rainbow.

Step 4: You should have seen a spectrum, but with no curve.

Step 5: If you were able to use the sunlight, you should have gotten the same results as when you used your flashlight. However, spectrums created with sunlight are much brighter than those formed with a flashlight.

Why?

White light (WIIT LIIT) is not really white at all! It's made up of red, orange, yellow, green, blue, indigo, violet and all the shades of color in between. When white light hits a prism in just the right way it is **refracted,** or bent. Bent just the right way, light separates into all its parts—and lets all the colors of the rainbow shine through.

Why aren't rainbows straight? In Step 1 you made a curved mini-rainbow with a glass of water. Because the surface of the water curved, when the water refracted the light, it made a curved spectrum. Rainbows in the sky are formed when sunlight is refracted through the curved surface of—raindrops!

Artists have found their own pot of gold, but it's not at the *end* of the rainbow, it's inside. Without the secret hidden in white light, artists wouldn't be nearly as colorful.

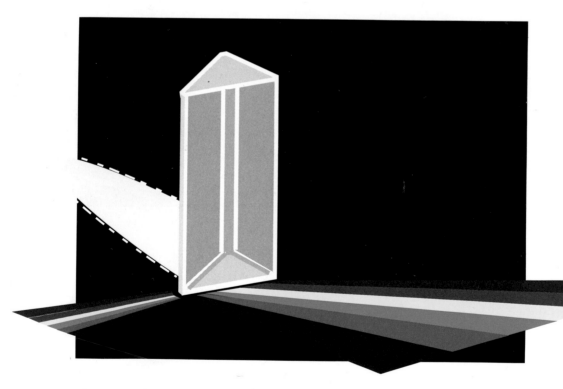

A prism refracts, or bends, light into the colors of a rainbow. All the colors that artists use would not be possible without the science of light—photology.

HOW CAN YOU MAKE SIDEWALK CHALK?

Are you a sidewalk artist? Do you use colored chalk to make drawings on the sidewalk or in the school yard?

Sidewalk artists depend on their tools—colored chalks—to create. However, their tools depend on the science of **chemistry** to work.

Step 1 Pour about 1 inch (2.5 cm) of clean sand into a clean soup can.

Smear petroleum jelly all over the inside of the toilet paper roll. Stand the toilet paper roll in the sand at the bottom of the can. Make sure the roll will stand upright on its own.

Step 2 Measure 1/2 cup (120 ml) of **plaster of Paris** (PLAS tuhr uhv PAYR iss) powder into another clean soup can. Add 4 teaspoons of powdered tempera paint—any color you choose—to the plaster of Paris. Stir the powders together with a knife.

Add 2 ounces or 1/4 cup (60 ml) of water to the plaster of Paris-tempera paint mixture. Stir until the mixture is smooth and all the same color.

What You Need

- ✓ plaster of Paris (available in hardware or craft stores)
- ✓ powdered tempera paints (available in craft or art supply stores)
- ✓ petroleum jelly
- ✓ empty toilet tissue or paper towel rolls
- ✓ clean sand
- ✓ 2 clean soup cans with tops removed
- ✓ teaspoon
- ✓ measuring cup
- ✓ knife
- ✓ tap water

Step 3 Pour the mixture carefully into the toilet paper roll, and let it sit until the next day. Then, carefully peel the toilet paper roll away. You should have a ready-to-use piece of sidewalk chalk!

Try to scratch it with your finger. Find a place to test your chalk. Does it work the way it's supposed to? Write all of your **observations** down in your **science log.** Take photographs of the sidewalk art you create for your final display.

Step 4 Repeat Step 1 through Step 3 using different colors of tempera paint, different combinations of colors, and different amounts of paint. What differences do you observe? Write everything down in your science log.

Is This What Happened?

Steps 3-4: You should have created a solid, chalky-looking material that you could scratch with your finger. When you tried to draw with it, it should have acted like any other piece of chalk.

Pour the plaster of Paris and tempera paint carefully into the toilet paper roll. Let it dry for a whole day before peeling off the paper roll.

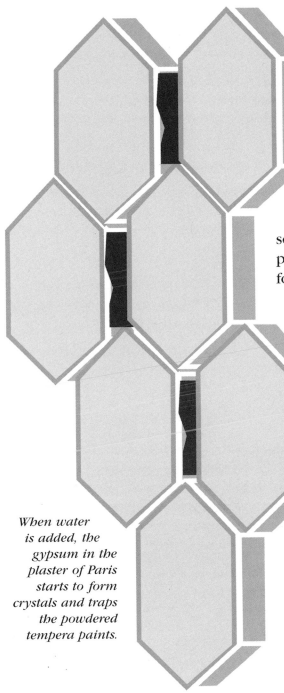

When water is added, the gypsum in the plaster of Paris starts to form crystals and traps the powdered tempera paints.

Why?

The two ingredients—plaster of Paris and powdered tempera paint—go through a chemical process to become chalk. Plaster of Paris is made from a **mineral** (MIN er uhl) called **gypsum** (JIP suhm). Gypsum is a soft, white solid. When plaster of Paris is in its powder form, the gypsum inside can't form into **crystals** (KRIS tuhlz). However, if you add water, the gypsum can and does form tiny crystals.

In Step 2, you mixed powdered tempera paint with the gypsum. When you poured water into the mixture, you formed a thick **suspension** (sus PEN shun) and the gypsum began to form crystals. As the plaster hardened, the particles of powdered paint were trapped between the gypsum crystals—and out came colored chalk.

Thanks to the science of chemistry, sidewalks everywhere can explode with the color and creativity of young artists like you.

HOW DOES ANIMATION WORK?

Watching Saturday morning cartoons or animated full-length feature films can be a lot of fun. Did you ever wonder how animators get their two-dimensional characters to act?

Animators actually draw thousands of pictures and put them together, but they can't make cartoons move. We do that on our own through **persistence of vision** (per SIST ens uhv VIH jhun).

What To Do

Step 1 Use a compass to draw a 6-inch (15-cm) circle on heavy cardboard. Be sure to make a hole in the center of the circle with your compass. Ask an adult to help you cut out the circle with a sharp knife.

Step 2 Cut a strip of posterboard 3 inches (7.5 cm) by 20 1/4 inches (51.4 cm). Use your ruler to draw a line along the long edge of the strip, 1 inch (2.5 cm) from the top.

Starting from the left, measure 1 9/16 inches (4 cm) from the edge and make a mark on your line. Then, measure 1 9/16 inches (4 cm) from the mark you just made and make another mark. Keep working across the entire strip until you have 12 marks.

What You Need
✓ corrugated cardboard or other heavy cardboard
✓ posterboard
✓ drawing compass
✓ pencil
✓ ruler
✓ white glue
✓ flat black spray paint
✓ sharp hobby knife
✓ large push pin with a plastic top
✓ wood rod, or dowel—3/4 inch (2 cm) by 6 inches (15 cm)
✓ an adult to help

Step 3 Ask an adult to help you make 12 slits in your strip. At each of the marks you just made, make a cut from the top of your strip straight down to your mark on the line. You should end up with 12 1-inch-long slits equally spaced along your strip.

Make another slit 1/16 inch (1.5 mm) to the right of each of the 12 slits you just made. Remove the 12 narrow strips between the slits by cutting between them along your line. You'll have 12 slots when you're finished—each 1 inch (2.5 cm) tall by 1/16 inch (1.5 mm) wide.

Step 4 Glue the bottom of the strip to the circle you cut out in Step 1. Use white glue all along the rim of the circle and where the two edges of the strip overlap. Hold the strip in place until the glue sets—and you have an **animation** (an i MAY shun) drum.

Ask an adult to help you use flat black spray paint to paint the inside and outside of the drum. Be sure to read all the directions and caution labels on your paint can. After it drys, push the large push pin through the hole in the circle into the wood dowel. Holding the dowel "handle," make sure the circle can spin. If it can't, pull out the push pin a little. You've just created a **zoetrope** (ZOH eh trohp)!

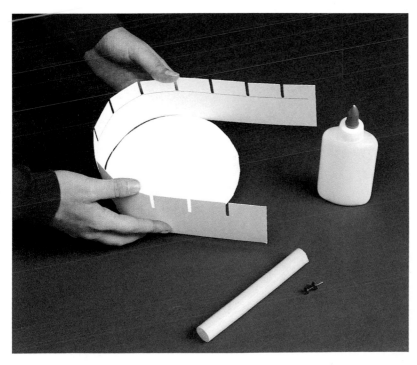

Glue the bottom strip to the circle, slots up. The strip will overlap. Glue those edges together, too.

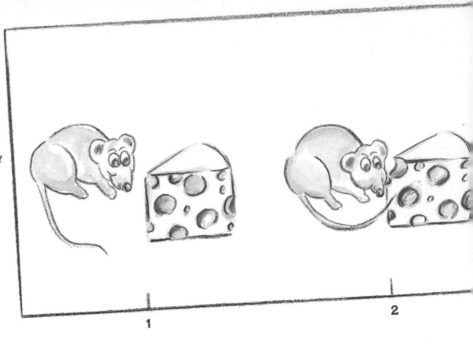

Draw 12 pictures centered over your marks. Each picture should be just a little different than the one before to show the next stage of the action. (Posterboard strip is shown larger than actual size.)

Step 5 Cut a strip of posterboard 1 7/8 inches (4.8 cm) by 18 1/2 inches (47 cm). Lay the strip lengthwise in front of you. Starting from the left, measure 3/4 inch (2 cm) from the edge and make a small pencil mark on the bottom of the strip.

Then, measure 1 9/16 inches (4 cm) from this mark and make another mark. Continue to make marks 1 9/16 inches (4 cm) apart until you have 12 marks.

Step 6 These marks show where the center of your drawings need to be. In the spaces above the marks, make a series of drawings that show action—a flower growing, ball bouncing, a stickman walking—anything you can imagine. You can create as many different strips as you like. Try to use color.

Step 7 Gently bend your finished picture strip by running the picture side over the edge of a table or around your fingers. Then fit the strip inside the circle of your zoetrope so that the pictures face the center.

In a room with good light, hold the zoetrope about 6 inches (15 cm) away from your face and look through the slots. Spin the circle, and what do you see? Record all your **observations** in your **science log.** Try other strips you've created. Do you see any differences?

Is This What Happened?

Step 7: Looking through the slots, you should see a moving picture! Your individual pictures are blended into one continuous action.

3   4   5

Why?

When you look at something, its image forms on the **retina** (RE tin uh) of your eye. A signal moves from the retina to your brain and you "see." However, the retina does not work as fast as objects can move. The image on your retina stays for about 1/15 second. This effect is called *persistence of vision*.

When you spin your zoetrope and look through the slots, each picture forms an image on your retina. Each new image comes so quickly that the old image has not had time to fade totally from your retina. Each image blends into the next so what you see moves like a movie or cartoon—and you have created animation. However, this effect of motion is only an **illusion** (il OO jhun). You are really seeing only one picture at a time.

Now, with the help of your zoetrope, you can take your artwork and get moving!

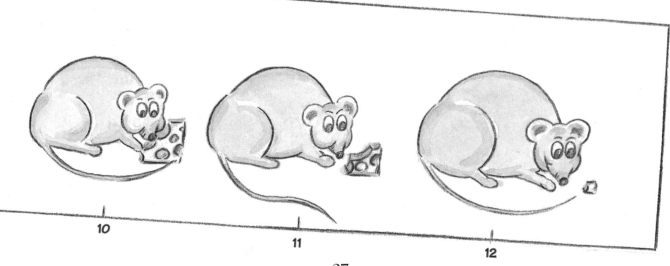

10   11   12

HOW TO DISPLAY YOUR PROJECT

When you finish your project, your teacher may ask you to share it with your class or show it at a science fair. Professional scientists often show their work to other people. Here are some tips on how to display your project.

Many students show their projects with a three-board, free-standing display. Before you start putting everything together, make a sketch of how you would like your display to look. This is the best time to make changes.

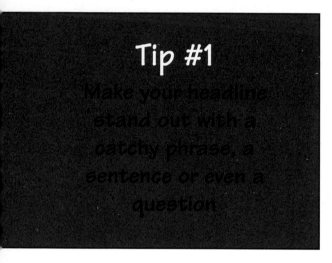

Tip #1

Make your headline stand out with a catchy phrase, a sentence or even a question

The title of your project should attract people's attention. It could be one or two words—a catchy phrase, a sentence or even a question. Use the largest lettering for your title. In your display, you should also state the scientific problem you were trying to solve. Use a question, like the chapter titles in this book, or state your problem in the form of a **hypothesis.**

If you have a computer of your own, or can use one at school, they work great for lettering. Or, you can neatly print on a white sheet of paper, and border your lettering with colored construction paper to make it stand out.

You'll also need to leave room to display the most important part of your project—your results. Show any photographs, drawings, charts, graphs or tables—anything that will help to explain what you've learned. You can use

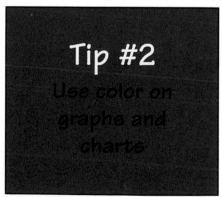

Tip #2

Use color on graphs and charts

black marker to make tables and charts, and colored marker for graphs. If you're handy with a computer, you might try to make your graphs and charts with a computer program!

Once you have all the pieces, tape everything into place. Follow the sketch you drew. Using tape will let you rearrange things until your display looks exactly how you want it. Then glue the pieces permanently.

As part of your exhibit, you'll want to include your **science log** and final report, along with any equipment you used, or models you made. Make sure your report is easy to read—neatly printed or typed. Be sure to label everything clearly.

Finally, you'll want to be able to tell people about your project. Practice what you want to say beforehand as many times as you can. Tell your parents, a friend, or even your dog about it. Then when a teacher or judge asks you about your project, you'll know what to say. You can share what you've discovered, and show that science really is part of everyone's life. After all, you've just become a real scientist!

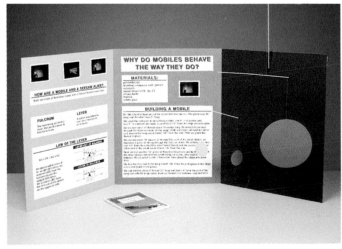

Sharing your results is an important part of being a scientist. A well-organized display will make explaining what you've learned easier.

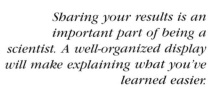

GLOSSARY

animation (an i MAY shun) – the process of drawing a series of pictures and showing them rapidly to create the illusion of movement

chemistry (KEM iss tree) – the study of matter—what things are made of; how it's classified, how it's put together, and the changes it goes through

chromatography (krohm uh TAH gruh fee) – separating a mixture into its colored parts

crystals (KRIS tuhlz) – solids found in nature that have a specific shape with a repeating pattern of sides and angles

diameter (dii A mi tuhr) – distance across a circle through the center, from rim to rim

dowel (DOW uhl) – a wooden rod

dyes (DIIZ) – colored substances used to make inks and coloring for fabric

fulcrum (FUL kruhm) – the balancing point of a lever; the point a lever is free to turn on

gypsum (JIP suhm) – a soft, white mineral used to make plaster of Paris

hypothesis (hii POTH uh siss) – a possible answer to a scientific question; sometimes called an educated guess because scientists use what they *already* know to guess how the experiment will turn out

illusion (il OO jhun) – seeing something different than it really is, as in animation—seeing motion in a series of still pictures

lever (LEH vuhr) – a rod or board free to turn on a fixed point, or fulcrum

mineral (MIN er uhl) – an element or compound found in nature

observation (ahb zer VAY shun) – information gathered by carefully using your senses; seeing, hearing, touching, smelling and tasting

opaque (oh PAYK) – anything that does not allow light to pass through it

persistence of vision (per SIST ens uhv VIH shun) – an image formed on the retina that remains for a short time, even after the object causing the image has changed or moved

physical laws (FIZ i kul LAWZ) – rules that scientists have found that describe how nonliving things work

plaster of Paris (PLAS tuhr uhv PAYR iss) – a powder made from gypsum, used to make plaster

principle (PRIN sih puhl) – laws, theories and concepts of science

prism (PRIZM) – a triangular piece of glass that refracts, or bends, light

reflect (ree FLEKT) – to bounce light off of a surface, like a mirror

refract (ree FRAKT) – to bend light as it passes through a transparent material

retina (RE tin ah) – a "screen" of light-sensitive cells at the back of the eye where the images we see are formed

science log (SII ens LAWG) – a notebook that includes the title of your project, the date you started, your list of materials, procedures you followed with dates and times, your observations and results

soluble (SAWL yoo buhl) – able to be dissolved into something

spectrum (SPEK trum) – the rainbow of colors produced when white light is refracted, or bent

suspension (sus PEN shun) – a mixture of one or more solids with a liquid; the solids are *not* dissolved, but are supported by the liquid

transparent (tranz PAYR ent) – allowing light to pass through so that you can see through an object or substance clearly

white light (WIIT LIIT) – light from the sun that allows us to see all colors

zoetrope (ZOH eh trohp) – a slotted drum free to turn on a handle that produces the illusion of moving pictures